recipes for homemade shakes, punches, hot drinks, and more

from the maker of SPLENDA® *Sweeteners*

PHOTOGRAPHS BY STEPHEN HAMILTON

CHRONICLE BOOKS
SAN FRANCISCO

Recipes on pages 33, 46, 49, 58, 70, 82, 102, 105, 109, 110, 114, and 117 provided courtesy of Kraft Food Holdings.

Design by Level, Calistoga, CA
Photographer: Stephen Hamilton
Photographer's Assistant: Matt Savage
Prop Stylists: Kelly McKaig & Paula Walters
Food Stylist: Josephine Orba
Assistant Food Stylist: Jill Kaczanowski

Library of Congress Cataloging-in-Publication Data available.

ISBN: 978-0-8118-6204-2

Manufactured in China.

10 9 8 7 6 5 4 3 2 1

Chronicle Books LLC
680 Second Street
San Francisco, California 94107

www.chroniclebooks.com

to all of you who have embraced a healthier lifestyle with the SPLENDA® Product Family, we dedicate this book and every sweet sip inside to you and your family. We raise our glasses in thanks for your continued support of the SPLENDA® Brand, and for helping us continue to create smiles with this sweet, delicious addition.

contents

INTRODUCTION 5

Substitution Chart 7

CHOCOLATE TREATS

Banana-Peanut Chocolate Smoothie. . . 11

French Vanilla White Hot Chocolate . . . 12

Iced Mocha Latte 15

Instant Hot Cocoa 16

Frozen Hot Chocolate 19

Hot Chocolate 20

WARMING COMFORTS

Hot Vanilla. 25

Vanilla Chai Latte 26

Mulled Cherry-Cranberry Warmer 29

Homemade Chai 30

Hot Brown Sugar Tea 33

Warm Tiramisù Latte 34

Hot Spiced Tea 37

SMOOTHIES AND SHAKES

Fuzzy Orange Smoothie. 41

Strawberry Smoothie 42

Lemon-Lime Milk Shake 45

Frozen Almond Cappuccino 46

Mango Yogurt Smoothie 49

Peach Melba Sipper 50

Creamsicle 53

Refreshing Summer Slushie 54

Malted Mocha Frappe 57

"Bananas Foster" 58

Banana-Raspberry Smoothie 61

COOLERS, SPARKLERS, AND ICED TEAS

Cantaloupe Agua Fresca 65

Fizzy Lemonade 66

Grapefruit-Raspberry Sparkler 69

Pear-Ginger Lemonade 70

Pomegranate Punch 73

Citrus Berry Spritzer 74

Apple Breeze 77

Vanilla-Orange Yogurt Float. 78

Raspberry Whip 81

Southern Iced Tea 82

Chamomile-Pomegranate Iced Tea 85

Citrus-Mint Iced Tea 86

Watermelon Lemonade 89

PARTY PUNCHES

Elegant Eggnog 93

Holiday Spiced Tea 94

Lemonade by the Pitcher! 97

Orange-Berry Sparkler. 98

Paradise Punch. 101

Peach-Flavored Green Tea Punch 102

Pineapple-Strawberry Punch 105

Sunshine Punch 106

Raspberry Tea Punch. 109

Tropical Pitcher Punch 110

Tropical Mango Punch. 113

Mulled Cider for a Crowd 114

Hot Cranberry Apple Cider. 117

ACKNOWLEDGMENTS 118

INDEX. 118

TABLE OF EQUIVALENTS 120

introduction

A favorite beverage can do much more than replenish. You can curl up in front of a fire with a warm cup of mulled cider and relax, savoring it sip by sip. On a hot afternoon or after a workout, invigorate yourself with a tumbler of chilled iced tea. Lift a glass of fruity punch to celebrate a special occasion. Indulge once in a while with a rich chocolate beverage, or satisfy your hunger with a thick smoothie.

The beverages in this book, *The SPLENDA® World of Sweet Drinks*, are perfect for all of these scenarios and countless others. But because they are made with SPLENDA® Sweeteners, they satisfy, refresh, and comfort with fewer calories from sugar than the traditional versions. Many of the recipes also have additional improvements, such as a reduction in fat.

The beverages have been grouped into sections that easily allow you to find just the right drink to suit your mood or the occasion. Each recipe is accompanied with nutritional information per serving. Some recipes are higher in calories than others, so if you are watching calories, be sure to check the nutritional information per serving to help you keep track.

So, no matter what the occasion, when thirst strikes, you'll find a beverage here to refresh, restore, and replenish. Cheers!

UNDERSTANDING SPLENDA® BRAND SWEETENER (SUCRALOSE)

SPLENDA® Brand Sweetener (sucralose) is produced by a patented multistep process that starts with sugar. SPLENDA® Brand Sweetener is suitable for the whole family. Here is some helpful information about the SPLENDA® Sweetener Products used in this book.

About SPLENDA® No Calorie Sweetener

SPLENDA® No Calorie Sweetener is a great alternative to sugar. It gets its sweetness from sucralose (SPLENDA® Brand Sweetener) and comes in many innovative forms. The familiar pastel yellow SPLENDA® No Calorie Sweetener Packets are premeasured for sweetening beverages and sprinkling on food. The texture of the sweetener is fine and powdery. Each packet provides the sweetness of two teaspoons of sugar. Use them to sweeten smoothies, shakes, and hot drinks.

SPLENDA® No Calorie Sweetener, Granulated, has a coarser texture than the SPLENDA® Packets and measures and pours just like sugar. Therefore, you would use one cup of SPLENDA® Granulated Sweetener to replace one cup of sugar in a recipe. This product works well to sweeten a large-batch beverage recipe, such as a pitcher of lemonade or a blender full of smoothies. When substituting SPLENDA® Granulated Sweetener for sugar in beverages, no adjustments are needed—just add it to the liquid and stir until the SPLENDA® dissolves.

If you plan to use SPLENDA® Granulated Sweetener in sugar-based recipes for baked goods, you should go to www.splenda.com for tips on how to achieve optimal results.

SPLENDA® Flavors for Coffee are a blend of SPLENDA® No Calorie Sweetener and popular coffeehouse flavors: French Vanilla, Mocha, Cinnamon Spice, Caramel, and Hazelnut. Each packet provides the sweetness of two teaspoons of sugar along with delicious coffeehouse flavor. Now you can enjoy a gourmet-flavored cup of coffee without all of the extra calories from sugar.

SPLENDA® No Calorie Sweetener FLAVOR ACCENTS™ Packets are designed specifically for cold drinks, and will perk up your bottled water or iced tea with fruity flavors like lemon and raspberry. An individual packet will add a mild fruit taste to 8 fluid ounces of beverage; use two packets for a 16-fluid-ounce serving.

About SPLENDA® Brown Sugar Blend

When you want the rich flavor of brown sugar, but with half the calories and carbohydrates, use SPLENDA® Brown Sugar Blend. This is a unique blend of brown sugar and SPLENDA® Brand Sweetener (sucralose). Because it is twice as sweet as regular brown sugar, you only need to use half as much. So, if your recipe calls for a cup of brown sugar, you only need a half-cup of SPLENDA® Brown Sugar Blend. Other than measuring, no other adjustments need to be made when cooking with SPLENDA® Brown Sugar Blend. It does contain calories and carbohydrates that should be taken into account when planning meals.

substitution chart

FOR SPLENDA® Sweeteners AND SUGAR

SUGAR:	SPLENDA® NO CALORIE SWEETENER, GRANULATED:	SPLENDA® SUGAR BLEND OR SPLENDA® BROWN SUGAR BLEND:
2 teaspoons	2 teaspoons (or 1 packet)	1 teaspoon
¾ cup	¾ cup	6 tablespoons
⅔ cup	⅔ cup	⅓ cup (or 5 tablespoons plus 1 teaspoon)
½ cup	½ cup	¼ cup (or 4 tablespoons)
⅓ cup	⅓ cup	2 tablespoons plus 2 teaspoons (or 8 teaspoons)
¼ cup	¼ cup	⅛ cup (or 2 tablespoons)

EXOTIC AND FAMILIAR at the same time, chocolate never fails to please. Here's a collection of surefire liquid pleasures, both hot and cold. Most of these recipes use unsweetened cocoa powder, which is available in two types. Natural cocoa powder, which is basically ground cacao beans, is the familiar supermarket variety. Dutch-process cocoa has been treated with alkali, which darkens the color and mellows the flavor. In beverages, the two are interchangeable, although drinks made with Dutch-process cocoa will look darker and may taste a bit richer. If the labeling isn't clear on the cocoa package, check the ingredients panel for some form of the word "alkali."

chocolate treats

MAKES 4 SERVINGS | Preparation Time: 5 minutes | Freezing Time: 10 minutes

banana-peanut chocolate smoothie

This satisfying smoothie is great for a breakfast on the go.

- 1 small, ripe banana, peeled and sliced
- 1 cup crushed ice
- ½ cup nonfat sour cream
- 3 tablespoons reduced-fat peanut butter
- 4 packets SPLENDA® No Calorie Sweetener
- 2 teaspoons unsweetened cocoa powder
- ½ teaspoon imitation banana extract (optional)

1. Freeze the sliced banana until slightly firm, about 10 minutes.
2. Combine the frozen banana, ice, sour cream, peanut butter, SPLENDA® No Calorie Sweetener, cocoa, and banana extract, if using, in a blender. Process the mixture, stopping to scrape down the sides of the blender as needed, until smooth.
3. Pour into glasses and serve immediately.

NUTRITIONAL INFORMATION

Serving Size: 4 fl oz
Total Calories: 110
Calories from Fat: 40
Total Fat: 4 g
Saturated Fat: 1 g
Cholesterol: 0 mg
Sodium: 100 mg
Total Carbohydrates: 15 g
Dietary Fiber: 2 g
Sugars: 8 g
Protein: 6 g

EXCHANGES PER SERVING
1 starch, 1 fat

MAKES 1 SERVING | Preparation Time: 2 minutes

french vanilla white hot chocolate

Sipping a smooth, vanilla-y white hot chocolate is a great way to treat yourself.

- **½ cup fat-free half-and-half**
- **½ cup skim milk**
- **1 packet SPLENDA® Flavors for Coffee, French Vanilla**
- **2 tablespoons white chocolate morsels**

Combine the half-and-half, skim milk, SPLENDA® French Vanilla Flavors for Coffee, and white chocolate morsels in a large microwave-safe mug. Microwave on Medium, stirring every 30 seconds, until the milk is hot and the white chocolate is smooth and melted, about 1 minute. Serve hot.

NUTRITIONAL INFORMATION

Serving Size: 8 fl oz
Calories: 260
Calories from Fat: 100
Total Fat: 11 g
Saturated Fat: 6 g
Cholesterol: 15 mg
Sodium: 260 mg
Total Carbohydrates: 33 g
Dietary Fiber: 0 g
Sugars: 28 g
Protein: 9 g

EXCHANGES PER SERVING

2 starches, ½ reduced-fat milk, 1½ fats

MAKES 4 SERVINGS | Preparation Time: 10 minutes | Freezing Time: 8 hours

iced mocha latte

It's hard to beat this caffeinated treat for an afternoon pick-me-up.

⅔ cup SPLENDA® No Calorie Sweetener, Granulated

2 tablespoons unsweetened cocoa powder, preferably Dutch-process

2 tablespoons instant coffee granules

2 cups boiling water

2 cups fat-free half-and-half, divided

Frozen whipped topping, thawed, and chocolate curls for garnish (optional)

1. Stir together the SPLENDA® Granulated Sweetener, cocoa powder, and instant coffee in a medium bowl. Gradually whisk in the boiling water. Stir in 1 cup of the half-and-half. Divide the coffee mixture among ice cube trays. Freeze until completely solid, at least 8 hours or overnight.
2. Pour the remaining 1 cup half-and-half into a blender. With the machine running, add the frozen coffee cubes, a few at a time, and blend until smooth.
3. Pour into glasses and garnish with the whipped topping and chocolate curls, if desired. Serve immediately.

NUTRITIONAL INFORMATION

Serving Size: 8 fl oz, without garnishes
Calories: 100
Calories from Fat: 0
Total Fat: 0 g
Saturated Fat: 0 g
Cholesterol: 0 mg
Sodium: 105 mg
Total Carbohydrates: 18 g
Dietary Fiber: <1 g
Sugars: 8 g
Protein: 5 g

EXCHANGES PER SERVING

1 starch

MAKES 22 SERVINGS | Preparation Time: 5 minutes

instant hot cocoa

Keep this big-batch easy-to-make mix on hand for when you want a quick cup of hot cocoa without the bother of heating up milk.

HOT COCOA MIX:

One 9.6-ounce box instant nonfat dry milk powder

1¼ cups SPLENDA® No Calorie Sweetener, Granulated

⅔ cup powdered nondairy coffee creamer

⅔ cup Dutch-process cocoa powder

FOR EACH SERVING:

¼ cup Hot Cocoa Mix, above

¾ cup boiling water

1. To make the hot cocoa mix, combine the instant milk powder, SPLENDA® Granulated Sweetener, nondairy creamer, and cocoa in a large bowl. Transfer to an airtight container. Store the cocoa mix in a cool, dark place for up to 6 months.

2. For each serving of hot cocoa, stir ¼ cup of the hot cocoa mix with ¾ cup boiling water in a coffee mug. Serve immediately.

NOTE: Hot beverages will hold their heat best if served in warmed cups. Fill the cups with boiling water or very hot tap water and let stand while making the beverage. Discard the water and quickly dry the cups before filling and serving.

NUTRITIONAL INFORMATION

Serving Size: 6 fl oz
Calories: 70
Calories from Fat: 15
Total Fat: 1 g
Saturated Fat: 1 g
Cholesterol: 0 mg
Sodium: 70 mg
Total Carbohydrates: 9 g
Dietary Fiber: 1 g
Sugars: 8 g
Protein: 5 g

EXCHANGES PER SERVING

½ starch, ½ fat

MAKES 2 SERVINGS | Preparation Time: 5 minutes

frozen hot chocolate

The combination of cocoa and chocolate chips gives this blended beverage lots of rich flavor.

1½ cups ice cubes

¼ cup SPLENDA® No Calorie Sweetener, Granulated

¼ cup unsweetened cocoa powder

¼ cup semisweet chocolate chips

2 tablespoons instant nonfat dry milk powder

1 cup 1% low-fat milk

½ teaspoon vanilla extract

1. In the order listed, place the ice cubes, SPLENDA® Granulated Sweetener, cocoa powder, chocolate chips, instant milk powder, low-fat milk, and vanilla in a blender. Process the mixture, stopping to scrape down the sides of the blender as needed, until smooth.

2. Pour into glasses and serve immediately.

NUTRITIONAL INFORMATION

Serving Size: 12 fl oz
Calories: 200
Calories from Fat: 80
Total Fat: 9 g
Saturated Fat: 5 g
Cholesterol: 5 mg
Sodium: 95 mg
Total Carbohydrates: 27 g
Dietary Fiber: 5 g
Sugars: 20 g
Protein: 9 g

EXCHANGES PER SERVING

1½ starches, ½ low-fat milk, 1 fat

MAKES 2 SERVINGS | Preparation Time: 5 minutes | Cooking Time: 5 minutes

hot chocolate

Nothing takes away the chill like a cup of hot chocolate . . .

8 packets SPLENDA® No Calorie Sweetener

3 tablespoons Dutch-process cocoa powder

2 cups 2% reduced-fat milk

1. Mix together the SPLENDA® No Calorie Sweetener and cocoa in a small saucepan. Gradually whisk in the milk.
2. Cook over medium heat, stirring constantly, until heated through, about 5 minutes. Pour into cups and serve immediately.

NOTE: Hot beverages will hold their heat best if served in warmed cups. Fill the cups with boiling water or very hot tap water and let stand while making the beverage. Discard the water and quickly dry the cups before filling and serving.

NUTRITIONAL INFORMATION

Serving Size: 8 fl oz
Calories: 140
Calories from Fat: 50
Total Fat: 6 g
Saturated Fat: 4 g
Cholesterol: 20 mg
Sodium: 125 mg
Total Carbohydrates: 16 g
Dietary Fiber: 2 g
Sugars: 12 g
Protein: 10 g

EXCHANGES PER SERVING

1 reduced-fat milk

THESE ARE the drinks to warm a cold winter night, or to soothe the spirit on a busy afternoon. We've covered the possibilities with coffee, tea, juices, and hot milk beverages. To keep beverages hot, serve them in hot mugs or cups. Pour hot water (boiling is best, but very hot tap water will do) into the mugs and let them stand for a minute or two. (Or, fill a microwave-safe mug with water, microwave on High until the water boils, and let stand for 1 minute.) Pour out the water and dry the mugs.

warming comforts

MAKES 1 SERVING | Preparation Time: 10 minutes | Cooking Time: 2 minutes

hot vanilla

Hot chocolate has its fans, but why not try hot vanilla for a change?

¾ cup nonfat milk

2 teaspoons SPLENDA® No Calorie Sweetener, Granulated

¼ teaspoon vanilla extract

1. Heat the milk in a small saucepan over low heat just to below the simmering point. (Or, pour the milk into a microwave-safe mug and heat on High just until the milk is very hot.) Remove from the heat. Stir in the SPLENDA® Granulated Sweetener and the vanilla until the SPLENDA® is dissolved.

2. Pour into a mug and serve hot.

NUTRITIONAL INFORMATION

Serving Size: 6 fl oz
Calories: 70
Calories from Fat: 0
Total Fat: 0 g
Saturated Fat: 0 g
Cholesterol: 9 mg
Sodium: 95 mg
Total Carbohydrates: 9 g
Dietary Fiber: 0 g
Sugars: 9 g
Protein: 6 g

EXCHANGES PER SERVING

1 fat-free milk

MAKES 1 SERVING | Preparation Time: 6 minutes

vanilla chai latte

A gently spiced, vanilla-infused latte could be the ultimate bedtime treat.

1 cup 1% reduced-fat milk

2 spiced chai tea bags

2 packets SPLENDA® Flavors for Coffee, French Vanilla

Ground cinnamon for garnish (optional)

1. Pour the milk into a microwave-safe mug and add the tea bags. Heat on High until the milk is hot. Let the tea steep for 4 minutes. Remove the tea bags with a spoon, pressing the bags gently to release the flavor. Discard the tea bags.
2. Stir in SPLENDA® Flavors for Coffee packets. Sprinkle with cinnamon, if desired. Serve hot.

NUTRITIONAL INFORMATION

Serving Size: 8 fl oz
Calories: 100
Calories from Fat: 20
Total Fat: 2.5 g
Saturated Fat: 1.5 g
Cholesterol: 10 mg
Sodium: 115 mg

Total Carbohydrates: 15 g
Dietary Fiber: 0 g
Sugars: 12 g
Protein: 7 g

EXCHANGES PER SERVING

1 reduced-fat milk

MAKES 4 SERVINGS | Preparation Time: 5 minutes | Cooking Time: 15 minutes

mulled cherry-cranberry warmer

Apple cider isn't the only fruit juice that takes well to gently spicing—this relaxing beverage blends the flavors of white cranberry and cherry.

4 cups light white cranberry juice drink

¼ cup SPLENDA® No Calorie Sweetener, Granulated

One .13-ounce package Cherry Flavor KOOL-AID® Unsweetened Soft Drink Mix

Four 3-inch cinnamon sticks

8 whole cloves

Orange slices for garnish

1. Stir the cranberry juice, SPLENDA® Granulated Sweetener, and KOOL-AID® Unsweetened Soft Drink Mix in a medium saucepan until the SPLENDA® and KOOL-AID® dissolve. Add the cinnamon sticks and cloves.
2. Bring to a boil over high heat, stirring often. Reduce the heat to low and simmer, uncovered, for 15 minutes. Remove the spices with a slotted spoon. Discard the cloves, but reserve the cinnamon sticks for garnish.
3. Pour into cups and garnish with the cinnamon sticks and orange slices. Serve hot.

NUTRITIONAL INFORMATION

Serving Size: 8 fl oz
Calories: 100
Calories from Fat: 5
Total Fat: 1 g
Saturated Fat: 0 g
Cholesterol: 0 mg
Sodium: 50 mg
Total Carbohydrates: 27 g
Dietary Fiber: 10 g
Sugars: 0 g
Protein: 1 g

EXCHANGES PER SERVING

2 fruits

MAKES 3 SERVINGS | Preparation Time: 10 minutes

homemade chai

There is no need to go to the tea shop when you can make fragrant chai at home.

- 2 cups water
- 3 slices peeled fresh ginger
- 3 packets SPLENDA® No Calorie Sweetener
- 12 black peppercorns
- 3 whole cloves
- 2 cardamom pods
- One 3-inch cinnamon stick
- 2 Darjeeling tea bags
- ½ cup 2% reduced-fat milk

1. Bring the water, ginger, SPLENDA® No Calorie Sweetener, peppercorns, cloves, cardamom, and cinnamon to a boil in a small saucepan over medium-high heat, stirring to dissolve the SPLENDA®. Remove from the heat and add the tea bags. Cover and let steep for 5 minutes. Remove the tea bags with a spoon, pressing the bags gently to release the flavor. Discard the tea bags and spices.
2. Add the milk and cook over medium heat until thoroughly heated. Do not boil. Pour into mugs and serve immediately.

NUTRITIONAL INFORMATION

Serving Size: 6½ fl oz
Calories: 40
Calories from Fat: 10
Total Fat: 1 g
Saturated Fat: 1 g
Cholesterol: 5 mg
Sodium: 30 mg
Total Carbohydrates: 7 g
Dietary Fiber: 2 g
Sugars: 2 g
Protein: 2 g

EXCHANGES PER SERVING

½ starch

MAKES 3 SERVINGS | Preparation Time: 10 minutes

hot brown sugar tea

You'll find a hint of caramel flavor in this delightful tea. Add a splash of cream or milk, if desired.

3 orange pekoe tea bags
3 cups boiling water
¼ cup SPLENDA® Brown Sugar Blend

1. Place the tea bags in a teapot. Add the boiling water and cover. Let steep for 5 minutes. Remove the tea bags with a spoon, pressing the bags gently to release the flavor. Discard the tea bags.

2. Add the SPLENDA® Brown Sugar Blend and stir well to dissolve. Pour into cups and serve hot.

NUTRITIONAL INFORMATION

Serving Size: 8 fl oz
Calories: 80
Calories from Fat: 0
Total Fat: 0 g
Saturated Fat: 0 g
Cholesterol: 0 mg
Sodium: 5 mg
Total Carbohydrates: 16 g
Dietary Fiber: 0 g
Sugars: 16 g
Protein: 0 g

EXCHANGES PER SERVING
1 starch

MAKES 1 SERVING | Preparation Time: 5 minutes

warm tiramisù latte

This luscious drink combines the flavors of one of the most popular Italian desserts. Warm, creamy, and sippable, it might be even better than tiramisù.

- ½ cup 1% reduced-fat milk
- ¼ cup brewed espresso coffee
- 1 packet SPLENDA® Flavors for Coffee, French Vanilla
- 1 packet SPLENDA® Flavors for Coffee, Cinnamon Spice
- 2 teaspoons mascarpone cheese
- ½ teaspoon unsweetened cocoa powder
- ⅛ teaspoon vanilla extract

1. Mix the milk, coffee, and both SPLENDA® Flavors for Coffee in a heatproof mug. Cook in a microwave on High, stirring occasionally, until the mixture is hot, 30 to 40 seconds.
2. Pour the milk mixture into a blender. Add the mascarpone, cocoa, and vanilla and top with the lid placed slightly ajar (see Note). Process on low speed until the mixture is frothy, about 15 seconds. Return to the mug and serve immediately.

NOTE: Placing the lid slightly ajar with allow the steam from the hot milk mixture to escape.

NUTRITIONAL INFORMATION

Serving Size: 7 fl oz
Calories: 150
Calories from Fat: 90
Total Fat: 10 g
Saturated Fat: 6 g
Cholesterol: 30 mg
Sodium: 75 mg
Total Carbohydrates: 9 g
Dietary Fiber: 0 g
Sugars: 8 g
Protein: 6 g

EXCHANGES PER SERVING
½ reduced-fat milk, 1½ fats

MAKES 1 SERVING | Preparation Time: 10 minutes

hot spiced tea

With very little effort, you can transform your cup of tea into an extraordinary aromatic treat.

1 orange pekoe tea bag

½ cinnamon stick

4 whole cloves

1 cup boiling water

3 packets SPLENDA® No Calorie Sweetener

1 tablespoon orange juice, preferably fresh

1 teaspoon fresh lemon juice

Orange or lemon slices for garnish (optional)

1. Put the tea bag, cinnamon, and cloves in a teapot. Add the boiling water and cover. Let steep for 5 minutes. Remove the tea bag with a spoon, gently pressing the bag to release the flavor. Discard the tea bag. Add the SPLENDA® Packets and the orange and lemon juices. Stir to dissolve the SPLENDA®.

2. Pour the tea through a sieve into a cup and serve hot. Garnish with orange or lemon slices if desired.

NUTRITIONAL INFORMATION

Serving Size: 8 fl oz
Calories: 25
Calories from Fat: 0
Total Fat: 0 g
Saturated Fat: 0g
Cholesterol: 0 mg
Sodium: 20 mg

Total Carbohydrates: 6 g
Dietary Fiber: 2 g
Sugars: 2 g
Protein: 0 g

EXCHANGES PER SERVING
½ starch

THICK AND SATISFYING, these blender drinks are designed to fill you up. Need a quick breakfast on the run or a snack to hold you over until supper? Try one of the fruit-based smoothies or slushies. The milk-shake-style drinks are all reduced in fat as well as in sugar, so get ready to slurp every last drop through your straw. For the best blended drinks with the smoothest texture, use a blender and not a food processor (the liquid ingredients can seep through the shaft hole in a processor, and make quite a mess on the counter). Stop the blender a couple of times while processing and scrape down the sides with a rubber spatula to more effectively incorporate the ingredients.

smoothies and shakes

MAKES 2 SERVINGS | Preparation Time: 5 minutes

fuzzy orange smoothie

Blend banana slices and frozen peaches with orange juice to make a particularly thick and satisfying smoothie.

- **1 small ripe banana, sliced**
- **1 cup frozen peach slices**
- **½ cup orange juice, preferably fresh**
- **2 tablespoons SPLENDA® No Calorie Sweetener, Granulated**
- **2 tablespoons fresh lime juice**
- **½ teaspoon vanilla extract**
- **¼ teaspoon ground cinnamon**

1. In the order listed, combine the banana, peach slices, orange juice, SPLENDA® Granulated Sweetener, lime juice, vanilla, and cinnamon in a blender. Process the mixture, stopping to scrape down the sides of the blender as needed, until smooth.
2. Pour into glasses and serve immediately.

NUTRITIONAL INFORMATION

Serving Size: 6 fl oz
Calories: 100
Calories from Fat: 0
Total Fat: 0 g
Saturated Fat: 0 g
Cholesterol: 0 mg
Sodium: 0 mg
Total Carbohydrates: 24 g
Dietary Fiber: 2 g
Sugars: 20 g
Protein: 1 g

EXCHANGES PER SERVING

1½ fruits

MAKES 3 SERVINGS | Preparation Time: 5 minutes

strawberry smoothie

Purchase frozen strawberries, or freeze fresh berries on a baking sheet, then store in zippered plastic bags in the freezer to make this popular smoothie.

2 cups frozen unsweetened strawberries

1 cup 1% low-fat milk

1 cup plain nonfat yogurt

½ cup SPLENDA® No Calorie Sweetener, Granulated

One .13-ounce package Strawberry Flavor KOOL-AID® Unsweetened Soft Drink Mix

1. Combine the strawberries, milk, yogurt, SPLENDA® Granulated Sweetener, and KOOL-AID® Unsweetened Soft Drink Mix in a blender. Process the mixture, stopping to scrape down the sides of the blender as needed, until smooth.
2. Pour into glasses and serve immediately.

NUTRITIONAL INFORMATION

Serving Size: 8 fl oz
Calories: 120
Calories from Fat: 10
Total Fat: 1 g
Saturated Fat: 1 g
Cholesterol: 5 mg
Sodium: 160 mg
Total Carbohydrates: 24 g
Dietary Fiber: 3 g
Sugars: 19 g
Protein: 7 g

EXCHANGES PER SERVING
1 fruit, ½ fat-free milk

MAKES 3 SERVINGS | Preparation Time: 5 minutes

lemon-lime milk shake

Go ahead and indulge in this delicious shake with its unexpected flavors of lemon and lime.

1 pint no-sugar-added light vanilla ice cream

1 cup 1% low-fat milk

¼ cup SPLENDA® No Calorie Sweetener, Granulated

One .13-ounce package Lemon-Lime Flavor KOOL-AID® Unsweetened Soft Drink Mix

1. Combine the ice cream, milk, SPLENDA® Granulated Sweetener, and KOOL-AID® Unsweetened Soft Drink Mix in a blender. Process the mixture, stopping to scrape down the sides of the blender as needed, until smooth.
2. Pour into glasses and serve immediately.

NUTRITIONAL INFORMATION

Serving Size: 6 fl oz
Calories: 180
Calories from Fat: 60
Total Fat: 7 g
Saturated Fat: 4 g
Cholesterol: 20 mg
Sodium: 140 mg
Total Carbohydrates: 22 g
Dietary Fiber: 0 g
Sugars: 12 g
Protein: 7 g

EXCHANGES PER SERVING

1½ starches, 1 fat

MAKES 4 SERVINGS | Preparation Time: 5 minutes

frozen almond cappuccino

This frosty coffee drink is just like one you might get at your favorite coffeehouse, but without all of the sugary calories.

2 cups ice cubes

¼ cup instant nonfat dry milk powder

⅓ cup SPLENDA® No Calorie Sweetener, Granulated

5 teaspoons instant espresso granules

1½ cups 1% low-fat milk

½ teaspoon almond extract

1. In the order listed, combine the ice cubes, instant milk powder, SPLENDA® Granulated Sweetener, espresso granules, milk, and almond extract in a blender. Process the mixture, stopping to scrape down the sides of the blender as needed, until smooth.
2. Pour into glasses and serve immediately.

NUTRITIONAL INFORMATION

Serving Size: 8 fl oz
Calories: 60
Calories from Fat: 10
Total Fat: 1 g
Saturated Fat: 1 g
Cholesterol: 5 mg
Sodium: 750 mg
Total Carbohydrates: 7 g
Dietary Fiber: 0 g
Sugars: 7 g
Protein: 5 g

EXCHANGES PER SERVING

½ starch

MAKES 3 SERVINGS | Preparation Time: 5 minutes

mango yogurt smoothie

Let the mango get good and ripe (it should yield when gently pressed) before making this smoothie.

1 cup ice cubes

1 large ripe mango, peeled, pitted, and chopped

¼ cup SPLENDA® No Calorie Sweetener, Granulated

1 cup plain nonfat yogurt

½ cup mango nectar or fresh orange juice

¼ teaspoon almond extract

Pinch of salt

1. In the order listed, combine the ice cubes, mango, SPLENDA® Granulated Sweetener, yogurt, mango nectar, almond extract, and salt in a blender. Process the mixture, stopping to scrape down the sides of the blender as needed, until smooth.
2. Pour into glasses and serve immediately.

NUTRITIONAL INFORMATION

Serving Size: 8 fl oz
Calories: 110
Calories from Fat: 0
Total Fat: 0 g
Saturated Fat: 0 g
Cholesterol: 0 mg
Sodium: 180 mg
Total Carbohydrates: 24 g
Dietary Fiber: 2 g
Sugars: 20 g
Protein: 6 g

EXCHANGES PER SERVING

½ starch, 1 fruit

MAKES 5 SERVINGS | Preparation Time: 15 minutes | Chilling Time (for sauce): 2 hours

peach melba sipper

With frozen peaches and raspberries on hand, you can make this summertime beverage any time of the year.

RASPBERRY SAUCE

One 12-ounce package frozen unsweetened raspberries, thawed

¼ cup SPLENDA® No Calorie Sweetener, Granulated

1 tablespoon cornstarch

PEACH SMOOTHIES

One 16-ounce package frozen unsweetened sliced peaches

One 8-ounce container plain yogurt

¼ cup 2% reduced-fat milk or peach nectar

¼ cup SPLENDA® No Calorie Sweetener, Granulated

¼ teaspoon almond extract

1 cup ice cubes

NUTRITIONAL INFORMATION (WITH REDUCED-FAT MILK)

Serving Size: 8 fl oz
Calories: 160
Calories from Fat: 10
Total Fat: 1 g
Saturated Fat: 1 g
Cholesterol: 5 mg
Sodium: 45 mg
Total Carbohydrates: 34 g
Dietary Fiber: 3 g
Sugars: 27 g
Protein: 4 g

EXCHANGES PER SERVING

1½ starches, 1 fruit

1. To make the raspberry sauce, purée the raspberries in a food processor or blender. Using a rubber spatula, rub the mixture through a fine-meshed sieve over a bowl, discarding the seeds. Combine the SPLENDA® Granulated Sweetener and cornstarch in a small saucepan. Stir in the raspberry purée. Bring to a boil over medium heat, stirring constantly, then stir and boil for 1 minute. Pour into a bowl and cool. Cover and refrigerate until chilled, about 2 hours.

2. To make the smoothies, combine the peaches, yogurt, milk, SPLENDA® Granulated Sweetener, almond extract, and ice cubes in a food processor or blender. Process the mixture, stopping to scrape down the sides of the blender as needed, until smooth.

3. Spoon or drizzle 1 tablespoon of the raspberry sauce into each of 5 glasses. Pour in the peach smoothie mixture. Top each with the remaining raspberry sauce. Serve immediately.

MAKES 1 SERVING | Preparation Time: 5 minutes

creamsicle

The creamy orange flavor of this chilly drink will revive you on a warm afternoon.

⅓ cup orange juice, preferably fresh

¼ cup sugar-free nondairy whipped topping

2 tablespoons 1% reduced-fat milk

2 packets SPLENDA® Flavors for Coffee, French Vanilla

Combine the orange juice, whipped topping, milk, and the SPLENDA® Flavors for Coffee in a tall serving glass. Stir until the SPLENDA® Flavors for Coffee is dissolved. Add ice to fill the glass and serve.

NUTRITIONAL INFORMATION

Serving Size: About 6 fl oz
Calories: 100
Calories from Fat: 25
Total Fat: 2.5 g
Saturated Fat: 2 g
Cholesterol: 0 mg
Sodium: 15 mg
Total Carbohydrates: 18 g
Dietary Fiber: 0 g
Sugars: 3 g
Protein: 2 g

EXCHANGES PER SERVING
1 starch, ½ fat

MAKES 5 SERVINGS, PLUS FROZEN LEMONADE CUBES FOR ANOTHER USE | Preparation Time: 10 minutes | Freezing Time: 4 hours

refreshing summer slushie

Frozen lemonade cubes blended with your favorite soft drink make a cooling slushie that can help invigorate you on a hot afternoon. Use the leftover lemonade cubes to chill iced tea or other cold beverages.

LEMONADE CUBES

8 cups water

1 cup SPLENDA® No Calorie Sweetener, Granulated

One .13-ounce package Lemonade Flavor KOOL-AID® Unsweetened Soft Drink Mix

SLUSHIES

1½ cups water

½ cup SPLENDA® No Calorie Sweetener, Granulated

1 package of your favorite flavor KOOL-AID® Unsweetened Soft Drink Mix (package weight may vary slightly by flavor)

1½ cups frozen Lemonade Cubes, above

1. To make the lemonade cubes, stir together the water, SPLENDA® Granulated Sweetener, and KOOL-AID® Unsweetened Soft Drink Mix in a pitcher until the SPLENDA® and KOOL-AID® dissolve. Divide the lemonade among ice trays. Freeze until completely solid, at least 4 hours or overnight.
2. To make the slushies, combine the water, SPLENDA® Granulated Sweetener, KOOL-AID® Unsweetened Soft Drink Mix, and frozen lemonade cubes in a blender. (Store the leftover lemonade cubes in the freezer to use in iced tea and other beverages.) Process the mixture, stopping to scrape down the sides of the blender as needed, until smooth.
3. Pour into glasses and serve immediately.

NUTRITIONAL INFORMATION

Serving Size: 4 fl oz
Calories: 30
Calories from Fat: 0
Total Fat: 0 g
Saturated Fat: 0 g
Cholesterol: 0 mg
Sodium: 30 mg
Total Carbohydrates: 7 g
Dietary Fiber: 0 g
Sugars: 0 g
Protein: 0 g

EXCHANGES PER SERVING

½ starch

MAKES 5 SERVINGS | Preparation Time: 5 minutes

malted mocha frappe

In parts of New England, a coffee frappe (the "e" is silent) vies with other milk shakes for the most-popular ice cream beverage.

- 2 packets SPLENDA® Flavors for Coffee, Mocha
- ⅔ cup chocolate-flavored malt powder
- 2 tablespoons instant coffee granules
- 2 cups nonfat milk
- 2 cups light vanilla ice cream, preferably "slow churned"
- 1 cup crushed ice

1. In the order listed, combine the Mocha SPLENDA® Flavors for Coffee, malt powder, instant coffee, milk, ice cream, and crushed ice in a blender. Process the mixture, stopping to scrape down the sides of the blender as needed, until smooth.
2. Pour into glasses and serve immediately.

NUTRITIONAL INFORMATION

Serving Size: 8 fl oz
Calories: 290
Calories from Fat: 40
Total Fat: 4.5 g
Saturated Fat: 2.5 g
Cholesterol: 15 mg
Sodium: 170 mg
Total Carbohydrates: 54 g
Dietary Fiber: 2 g
Sugars: 45 g
Protein: 8 g

EXCHANGES PER SERVING

3 starches, ½ reduced-fat milk

MAKES 1 SERVING | Preparation Time: 5 minutes

"bananas foster"

New Orleans is famous for its cuisine, and the warm dessert bananas Foster is one of its most beloved recipes. Here is a warm smoothie with its spirit.

¾ cup 1% reduced-fat milk

1 tablespoon orange juice, preferably fresh

2 packets SPLENDA® Flavors for Coffee, Caramel

¼ teaspoon banana extract

One 2-inch piece of peeled ripe banana, sliced

1. Combine the milk, orange juice, SPLENDA® Flavors for Coffee, and banana extract in a heatproof mug. Microwave on High, stirring occasionally, until hot, about 50 seconds.
2. Pour the milk mixture into a blender. Add the sliced banana and top with the lid placed slightly ajar (see Note). Process until smooth. Return to the mug and serve immediately.

NOTE: Placing the lid slightly ajar will allow the steam from the hot milk mixture to escape.

NUTRITIONAL INFORMATION

Serving Size: About 7 fl oz
Calories: 110
Calories from Fat: 35
Total Fat: 3.5 g
Saturated Fat: 2.5 g
Cholesterol: 15 mg
Sodium: 75 mg
Total Carbohydrates: 12 g
Dietary Fiber: 0 g
Sugars: 10 g
Protein: 6 g

EXCHANGES PER SERVING

1 reduced-fat milk

MAKES 2 SERVINGS | Preparation Time: 5 minutes | Freezing Time: 10 minutes

banana-raspberry smoothie

The secret to a flavorful banana smoothie is nicely ripened fruit with lots of brown spots on the skin—avoid the overripe ones.

1 large ripe banana, peeled and sliced
1¼ cups frozen unsweetened raspberries
½ cup 1% low-fat milk
5 packets SPLENDA® No Calorie Sweetener

1. Place the banana slices on a baking sheet and freeze until partially frozen, about 10 minutes.
2. Place the frozen banana, raspberries, milk, and SPLENDA® No Calorie Sweetener in a blender. Process the mixture, stopping to scrape down the sides of the blender as needed, until smooth.
3. Pour into glasses and serve immediately.

NUTRITIONAL INFORMATION

Serving Size: 8 fl oz
Calories: 120
Calories from Fat: 10
Total Fat: 1 g
Saturated Fat: 1 g
Cholesterol: 0 mg
Sodium: 35 mg
Total Carbohydrates: 26 g
Dietary Fiber: 4 g
Sugars: 19 g
Protein: 4 g

EXCHANGES PER SERVING
2 fruits

REFRESHMENT is the key purpose of these cooling, light-bodied beverages. Look here for summertime thirst-quenchers to pour from an icy pitcher and sip in the cool shade of a tree. If you make your iced tea directly in the serving pitcher, be sure that the pitcher is heatproof. For a glass pitcher, it is safer to brew the tea in a separate container. You can bring the water for the tea to a boil in a saucepan, remove from the heat, add the tea bags, and steep right in the saucepan. Press the tea bags gently with a spoon to extract more flavor before discarding the bags.

coolers, sparklers, and iced teas

cantaloupe agua fresca

For the best results, make this Mexican thirst-quencher with very ripe melon.

½ cantaloupe, peeled, seeded, and cubed

4 cups cold water, divided

⅓ cup SPLENDA® No Calorie Sweetener, Granulated

2 tablespoons fresh lime juice

Cantaloupe wedges or lime slices for garnish (optional)

1. Process the cantaloupe and 1 cup of the water in a blender to make a coarse pulp. Pour into a large pitcher. Add the remaining 3 cups water, the SPLENDA® Granulated Sweetener, and the lime juice, and stir to dissolve the SPLENDA®.
2. Pour over ice in glasses and serve immediately, garnished with cantaloupe or lime, if desired.

NUTRITIONAL INFORMATION

Serving Size: 8 fl oz
Calories: 25
Calories from Fat: 0
Total Fat: 0 g
Saturated Fat: 0 g
Cholesterol: 0 mg
Sodium: 10 mg
Total Carbohydrates: 6 g
Dietary Fiber: 1 g
Sugars: 5 g
Protein: 1 g

EXCHANGES PER SERVING

½ starch

MAKES 8 SERVINGS | Preparation Time: 5 minutes

fizzy lemonade

Use club soda instead of plain water to add sparkle to lemonade.

- 1 cup SPLENDA® No Calorie Sweetener, Granulated
- One .13-ounce package Lemonade Flavor KOOL-AID® Unsweetened Soft Drink Mix
- 2 liters club soda, chilled
- Lemon slices for garnish (optional)

1. Mix the SPLENDA® Granulated Sweetener, KOOL-AID® Unsweetened Soft Drink Mix, and club soda in a large pitcher, stirring until the SPLENDA® and KOOL-AID® dissolve.
2. Pour into glasses over ice. Garnish with lemon slices, if desired. Serve chilled.

NUTRITIONAL INFORMATION

Serving Size: 8 fl oz
Calories: 0
Calories from Fat: 0
Total Fat: 0 g
Saturated Fat: 0 g
Cholesterol: 0 g
Sodium: 60 mg
Total Carbohydrates: 0 g
Dietary Fiber: 0 g
Sugars: 0 g
Protein: 0 g

EXCHANGES PER SERVING
Free

grapefruit-raspberry sparkler

Sweet and tart in just the right proportions, this drink would be welcome at any brunch.

- 1 cup fresh raspberries or frozen unsweetened raspberries, thawed
- 1¼ cups grapefruit juice, preferably fresh
- 3 tablespoons SPLENDA® No Calorie Sweetener, Granulated
- 1⅓ cups diet lemon-lime soda

1. Process the raspberries, grapefruit juice, and SPLENDA® Granulated Sweetener in a blender until smooth. Using a rubber spatula, rub the mixture through a sieve into a pitcher, discarding the solids.
2. Fill 4 tall glasses with ice. For each serving, add ½ cup of the raspberry mixture to the glass, then pour in ⅓ cup of the soda. Serve immediately.

NUTRITIONAL INFORMATION

Serving Size: 12 fl oz
Calories: 45
Calories from Fat: 0
Total Fat: 0 g
Saturated Fat: 0 g
Cholesterol: 0 mg
Sodium: 10 mg
Total Carbohydrates: 11 g
Dietary Fiber: 2 g
Sugars: 8 g
Protein: 1 g

EXCHANGES PER SERVING

½ fruit

MAKES 4 SERVINGS | Preparation Time: 5 minutes

pear-ginger lemonade

Not your typical lemonade, this one is made with pear nectar and an unexpected fillip of ginger.

3 cups pear nectar, chilled

½ cup fresh lemon juice

⅓ cup SPLENDA® No Calorie Sweetener, Granulated

4 teaspoons ginger juice (see Note)

1. Stir the pear nectar, lemon juice, SPLENDA® Granulated Sweetener, and ginger juice in a pitcher until the SPLENDA® dissolves.
2. Pour into glasses over ice and serve immediately.

NOTE: Bottled ginger juice can be purchased at specialty foods stores. To make your own, shred a piece of unpeeled fresh ginger about 5 inches long on the large holes of a box grater. Squeeze the shredded ginger in your fist over a bowl to extract the juice.

NUTRITIONAL INFORMATION

Serving Size: 6 fl oz
Calories: 120
Calories from Fat: 0
Total Fat: 0 g
Saturated Fat: 0 g
Cholesterol: 0 mg
Sodium: 10 mg
Total Carbohydrates: 32 g
Dietary Fiber: 1 g
Sugars: 29 g
Protein: 0 g

EXCHANGES PER SERVING
2 fruits

MAKES 6 SERVINGS | Preparation Time: 10 minutes | Freezing Time: 4 hours

pomegranate punch

Antioxidant-rich pomegranate juice is a delicious base for punch; make a double batch for a big party.

2 cups orange juice, preferably fresh

3 cups pomegranate juice, chilled

½ cup SPLENDA® No Calorie Sweetener, Granulated

2 tablespoons fresh lime juice

2 cups sparkling water, chilled

1. Divide the orange juice among ice cube trays. Freeze until completely solid, at least 4 hours or overnight.
2. Combine the pomegranate juice, SPLENDA® Granulated Sweetener, and lime juice in a large pitcher and stir until the SPLENDA® dissolves. Add the orange juice cubes and stir until the cubes begin to melt. Stir in the sparkling water.
3. Pour into glasses and serve immediately.

NUTRITIONAL INFORMATION

Serving Size: 6 fl oz
Calories: 120
Calories from Fat: 0
Total Fat: 0 g
Saturated Fat: 0 g
Cholesterol: 0 mg
Sodium: 10 mg
Total Carbohydrates: 30 g
Dietary Fiber: 0 g
Sugars: 25 g
Protein: 1g

EXCHANGES PER SERVING

2 fruits

MAKES 1 SERVING | Preparation Time: 5 minutes

citrus berry spritzer

This fizzy combination of orange, lemon, and raspberry flavors gets a colorful zing from raspberry juice concentrate drizzled on top.

½ cup seltzer or sparkling mineral water

1 packet SPLENDA® No Calorie Sweetener FLAVOR ACCENTS™, Lemon

1 packet SPLENDA® No Calorie Sweetener FLAVOR ACCENTS™, Raspberry

½ cup orange juice, preferably fresh

2 teaspoons thawed frozen raspberry or apple-raspberry juice concentrate (optional)

1. Combine the seltzer and Lemon and Raspberry SPLENDA® FLAVOR ACCENTS™ Packets in a tall glass and stir well. Add the orange juice. Fill the glass with ice.
2. If using, drizzle the raspberry juice concentrate over the top of the drink. Serve immediately.

NUTRITIONAL INFORMATION

Serving Size: About 8 fl oz
Calories: 80
Calories from Fat: 0
Total Fat: 0 g
Saturated Fat: 0 g
Cholesterol: 0 mg
Sodium: 10 mg
Total Carbohydrates: 18 g
Dietary Fiber: 0 g
Sugars: 14 g
Protein: 1 g

EXCHANGES PER SERVING

½ starch, 1 fruit

MAKES 1 SERVING | Preparation Time: 1 minute

apple breeze

The scent of cinnamon infuses this cool drink, one that is perfect for autumn entertaining.

- **⅔ cup sparkling apple cider**
- **1 tablespoon grapefruit juice**
- **1 packet SPLENDA® Flavors for Coffee, Cinnamon Spice**

Combine the apple cider, grapefruit juice, and SPLENDA® Flavors for Coffee in a tall glass and stir well. Fill the glass with ice and serve immediately.

NUTRITIONAL INFORMATION

Serving Size: About 6 fl oz
Calories: 95
Calories from Fat: 0
Total Fat: 0 g
Saturated Fat: 0 g
Cholesterol: 0 mg
Sodium: 0 mg
Total Carbohydrates: 24 g
Dietary Fiber: 0 g
Sugars: 21 g
Protein: 1 g

EXCHANGES PER SERVING

1 starch, 1½ fruit

MAKES 4 SERVINGS | Preparation Time: 5 minutes

vanilla-orange yogurt float

You don't have to be a kid to love a float. This one is fun to make and equally fun to drink.

¾ cup SPLENDA® No Calorie Sweetener, Granulated

One .13-ounce package Orange Flavor KOOL-AID® Unsweetened Soft Drink Mix

4 cups (1 liter) seltzer water, chilled (divided)

2 cups reduced-fat vanilla frozen yogurt, divided

1. Combine the SPLENDA® Granulated Sweetener, KOOL-AID® Unsweetened Soft Drink Mix, and 1 cup of the seltzer in a pitcher, stirring to dissolve the SPLENDA® and KOOL-AID®. Stir in the remaining 3 cups seltzer.
2. Pour 1 cup of the orange mixture into each of 4 tall glasses. Top each serving with a ½-cup scoop of the frozen yogurt. Serve immediately, with long spoons.

NUTRITIONAL INFORMATION

Serving Size: 12 fl oz
Calories: 220
Calories from Fat: 40
Total Fat: 4.5 g
Saturated Fat: 2.5 g
Cholesterol: 6.5 mg
Sodium: 75 mg
Total Carbohydrates: 36 g
Dietary Fiber: 0 g
Sugars: 21 g
Protein: 9 g

EXCHANGES PER SERVING

2½ starches, 1 fat

MAKES 1 SERVING | Preparation Time: 5 minutes

raspberry whip

When you crave the invigorating flavor of raspberry but don't have fresh ones in the house, you can still whip up this fruity dessert.

¼ cup sugar-free nondairy whipped topping

1 packet SPLENDA® No Calorie Sweetener FLAVOR ACCENTS™, Raspberry

1 small drop red food coloring

½ cup diet lemon-lime soda

1. Mix the whipped topping and SPLENDA® FLAVOR ACCENTS™ packet in a small bowl. Fold in the food coloring.
2. Fill a tall glass three-quarters full with ice. Pour in the soda, then top with the whipped topping mixture. Serve immediately.

NUTRITIONAL INFORMATION

Serving Size: About 6 fl oz
Calories: 40
Calories from Fat: 20
Total Fat: 2 g
Saturated Fat: 2 g
Cholesterol: 0 mg
Sodium: 10 mg
Total Carbohydrates: 6 g
Dietary Fiber: 0 g
Sugars: 0 g
Protein: 0 g

EXCHANGES PER SERVING
½ starch

MAKES 6 SERVINGS | Preparation Time: 15 minutes

southern iced tea

Many folks like their iced tea good and sweet, but you can reduce the amount of SPLENDA® No Calorie Sweetener if you wish.

5 cups water, divided

2 family-size orange pekoe tea bags

1 cup SPLENDA® No Calorie Sweetener, Granulated

¼ cup fresh lemon juice (optional)

Fresh mint sprigs and lemon slices for garnish (optional)

1. Bring 3 cups of the water to a boil in a medium saucepan over high heat. Remove from the heat. Add the tea bags and let stand for 10 minutes. Remove the tea bags with a spoon, pressing the bags gently to release the flavor. Discard the tea bags.

2. Add the SPLENDA® Granulated Sweetener and stir until dissolved. Stir in the remaining 2 cups water and the lemon juice, if using. Pour into a pitcher.

3. Pour into glasses over ice. Garnish with the mint sprigs and lemon slices, if desired. Serve chilled.

NUTRITIONAL INFORMATION

Serving Size: 6 fl oz
Calories: 15
Calories from Fat: 0
Total Fat: 0 g
Saturated Fat: 0 g
Cholesterol: 0 mg
Sodium: 0 mg
Total Carbohydrates: 4 g
Dietary Fiber: 0 g
Sugars: 0 g
Protein: 0 g

EXCHANGES PER SERVING

Free

MAKES 4 SERVINGS | Preparation Time: 5 minutes | Cooling Time: 1 hour

chamomile-pomegranate iced tea

Add tangy pomegranate juice to mellow chamomile tea for a zesty twist on iced tea.

4 chamomile tea bags

3 cups boiling water

1 cup pomegranate juice

⅓ cup SPLENDA® No Calorie Sweetener, Granulated

Fresh mint sprigs for garnish

1. Place the tea bags in a large heatproof measuring cup and add the boiling water. Let stand until cooled to room temperature, about 1 hour. Remove the tea bags with a spoon, pressing the bags gently to release the flavor. Discard the tea bags.
2. Add the pomegranate juice and SPLENDA® Granulated Sweetener and stir to dissolve the SPLENDA®.
3. Pour into glasses over ice, and garnish with the mint. Serve chilled.

NUTRITIONAL INFORMATION

Serving Size: 8 fl oz
Calories: 120
Calories from Fat: 0
Total Fat: 0 g
Saturated Fat: 0 g
Cholesterol: 0 mg
Sodium: 5 mg
Total Carbohydrates: 33 g
Dietary Fiber: 0 g
Sugars: 29 g
Protein: 0 g

EXCHANGES PER SERVING

½ fruit

citrus-mint iced tea

Look no further for a thirst-quenching iced tea to serve at your next summer cookout.

8 cups water, divided

5 orange pekoe tea bags

½ cup loosely packed fresh mint leaves

1 cup SPLENDA® No Calorie Sweetener, Granulated

1 cup orange juice, preferably fresh

⅓ cup fresh lemon juice

Orange slices, lemon slices, and fresh mint sprigs for garnish

1. Bring 2 cups of the water to a boil in a medium saucepan over high heat. Remove from the heat. Add the tea bags and mint leaves. Cover and let stand for 5 minutes. Remove the tea bags and mint with a spoon, pressing both gently to release the flavor. Discard the tea bags and mint. Add the remaining 6 cups cold water, the SPLENDA® Granulated Sweetener, and orange and lemon juices and stir until the SPLENDA® dissolves. Pour into a pitcher.

2. Serve in glasses over ice. Garnish with the orange slices, lemon slices, and mint sprigs. Serve chilled.

NUTRITIONAL INFORMATION

Serving Size: 8 fl oz
Calories: 15
Calories from Fat: 0
Total Fat: 0 g
Saturated Fat: 0 g
Cholesterol: 0 mg
Sodium: 5 mg
Total Carbohydrates: 4 g
Dietary Fiber: 0 g
Sugars: 3 g
Protein: 1 g

EXCHANGES PER SERVING
Free

MAKES 5 SERVINGS | Preparation Time: 15 minutes

watermelon lemonade

Watermelon and lemonade are both summertime classics, and here they are together as a cooling drink.

7 cups watermelon cubes, seeds removed

1 cup SPLENDA® No Calorie Sweetener, Granulated

1 cup fresh lemon juice

1. Process the watermelon in batches in a food processor until puréed. Press the mixture through a sieve into a pitcher, discarding the solids. You should have about 3 cups watermelon juice.
2. Add the SPLENDA® Granulated Sweetener and lemon juice and stir until the SPLENDA® dissolves. Pour into glasses over ice. Serve chilled.

NUTRITIONAL INFORMATION

Serving Size: 6 fl oz
Calories: 80
Calories from Fat: 0
Total Fat: 1 g
Saturated Fat: 0 g
Cholesterol: 0 mg
Sodium: 5 mg
Total Carbohydrates: 19 g
Dietary Fiber: 1 g
Sugars: 16 g
Protein: 2 g

EXCHANGES PER SERVING

1½ fruits

BRING OUT your punch bowl and fill it with one of these crowd-pleasers. These are big-batch drinks to serve a bunch of thirsty friends. To keep punch chilled without diluting it with ice, pour 2 cups of the punch into a pint container and freeze until solid. Unmold the frozen punch into the bowl, and it will keep the punch chilled as it melts. This trick also works with eggnog—use a pint of no-sugar-added vanilla ice cream to keep the eggnog cool—note that the addition of the ice cream adds calories that should be accounted for.

party punches

MAKES 15 SERVINGS | Preparation Time: 20 minutes | Chilling Time: 3 hours

elegant eggnog

It wouldn't be the holiday season without a cup of eggnog, and this recipe allows for indulgence without guilt.

1 cup SPLENDA® No Calorie Sweetener, Granulated

1 tablespoon cornstarch or arrowroot

1 teaspoon ground nutmeg

7 large egg yolks

4 cups whole milk

2 cups fat-free half-and-half

2 tablespoons vanilla extract or 1 teaspoon rum extract

1. Combine the SPLENDA® Granulated Sweetener, cornstarch, and nutmeg in a large, heavy saucepan. Whisk the egg yolks in a medium bowl until pale yellow. Scrape into the saucepan and whisk to blend with the dry ingredients. Gradually whisk in the milk. Cook over low heat, whisking constantly, until the mixture reads 175°F on an instant-read thermometer, about 7 minutes. Remove from heat and whisk in the half-and-half. Transfer to a large bowl and cool.

2. Cover with plastic wrap and refrigerate until chilled, at least 3 hours or up to 3 days.

3. Just before serving, stir in the vanilla extract. Ladle into punch cups and serve.

NUTRITIONAL INFORMATION

Serving Size: 4 fl oz
Calories: 90
Calories from Fat: 35
Total Fat: 4 g
Saturated Fat: 2 g
Cholesterol: 110 mg
Sodium: 70 mg
Total Carbohydrates: 7 g
Dietary Fiber: 0 g
Sugars: 5 g
Protein: 5 g

EXCHANGES PER SERVING

½ starch, 1 fat

MAKES 16 SERVINGS | Preparation Time: 10 minutes | Cooking Time: 25 minutes

holiday spiced tea

The welcoming aroma of orange, cinnamon, and cloves will greet your guests when you serve this tea at your holiday party.

- **12 cups water, divided**
- **Four 3-inch cinnamon sticks, plus more for garnish**
- **2 teaspoons whole cloves, plus more for garnish**
- **6 orange pekoe tea bags**
- **1¼ cups SPLENDA® No Calorie Sweetener, Granulated**
- **One 6-ounce can frozen orange juice concentrate, thawed**
- **¼ cup lemon juice, preferably fresh**
- **Orange and lemon slices for garnish**

1. Bring 4 cups of the water, the cinnamon sticks, and the whole cloves to a boil in a medium saucepan. Cover and reduce the heat to low. Simmer for 20 minutes.
2. When ready to serve, place the tea bags in a large heat-resistant bowl. Bring the remaining 8 cups water to a boil. Pour over the tea bags. Cover and let stand for 5 minutes. Remove the tea bags with a spoon, pressing the bags gently to release the flavor. Discard the tea bags. Add SPLENDA® Granulated Sweetener, orange juice concentrate, and lemon juice and stir to dissolve the SPLENDA®. Strain the spice mixture through a sieve into the tea mixture, discarding the solids.
3. Pour the hot tea into mugs and garnish with orange and lemon slices, cinnamon sticks, and cloves. Serve immediately.

NUTRITIONAL INFORMATION

Serving Size: 6 fl oz
Calories: 25
Calories from Fat: 0
Total Fat: 0 g
Saturated Fat: 0 g
Cholesterol: 0 mg
Sodium: 5 mg
Total Carbohydrates: 6 g
Dietary Fiber: 1 g
Sugars: 5 g
Protein: 0 g

EXCHANGES PER SERVING
½ fruit

MAKES 8 SERVINGS | Preparation Time: 5 minutes

lemonade by the pitcher!

You know that summer has arrived when you pour that first glass of lemonade . . .

5 cups water

1 cup fresh lemon juice

1 cup SPLENDA® No Calorie Sweetener, Granulated

Fresh mint sprigs and lemon slices for garnish (optional)

1. Stir the water, lemon juice, and SPLENDA® Granulated Sweetener in a large pitcher until the SPLENDA® dissolves.
2. Pour into glasses over ice, and garnish with the mint and lemon slices, if desired. Serve immediately.

NUTRITIONAL INFORMATION

Serving Size: 6 fl oz
Calories: 20
Calories from Fat: 0
Total Fat: 0 g
Saturated Fat: 0 g
Cholesterol: 0 mg
Sodium: 5 mg
Total Carbohydrates: 6 g
Dietary Fiber: 0 g
Sugars: 1 g
Protein: 0 g

EXCHANGES PER SERVING

½ starch

MAKES 16 SERVINGS | Preparation Time: 5 minutes

orange-berry sparkler

The festive color of this punch says "party," and it couldn't be easier to make.

2 cups SPLENDA® No Calorie Sweetener, Granulated

One .13-ounce package Orange Flavor KOOL-AID® Unsweetened Soft Drink Mix

One .13-ounce package Raspberry Flavor KOOL-AID® Unsweetened Soft Drink Mix

12 cups water

1 cup diet lemon-lime soda or diet ginger ale, chilled

Orange slices for garnish (optional)

1. Combine the SPLENDA® Granulated Sweetener, orange and raspberry KOOL-AID® Unsweetened Soft Drink Mixes, and water in a large pitcher, stirring to dissolve the SPLENDA® and KOOL-AID®.
2. Just before serving, stir in the lemon-lime soda. Pour into glasses over ice and garnish with orange slices, if desired. Serve chilled.

NUTRITIONAL INFORMATION

Serving Size: 8 fl oz
Calories: 5
Calories from Fat: 0
Total Fat: 0 g
Saturated Fat: 0 g
Cholesterol: 0 mg
Sodium: 25 mg
Total Carbohydrates: 2 g
Dietary Fiber: 0 g
Sugars: 2 g
Protein: 0 g

EXCHANGES PER SERVING

Free

MAKES 7 SERVINGS | Preparation Time: 5 minutes | Chilling Time: 2 hours

paradise punch

If your idea of paradise includes a tropical island laden with pineapples, then this punch is for you.

1 cup SPLENDA® No Calorie Sweetener, Granulated

One .19-ounce package Tropical Punch Flavor KOOL-AID® Unsweetened Soft Drink Mix

2 cups water

2 cups unsweetened pineapple juice

1 liter club soda, chilled

1. Combine the SPLENDA® Granulated Sweetener and the KOOL-AID® Unsweetened Soft Drink Mix in a punch bowl. Add the water and stir to dissolve the SPLENDA® and KOOL-AID®. Stir in the pineapple juice. Cover and refrigerate until chilled, at least 2 hours or overnight.
2. Just before serving, stir in the club soda. Serve over ice in tall glasses.

NUTRITIONAL INFORMATION

Serving Size: 8 fl oz
Calories: 40
Calories from Fat: 0
Total Fat: 0 g
Saturated Fat: 0 g
Cholesterol: 0 mg
Sodium: 50 mg
Total Carbohydrates: 10 g
Dietary Fiber: 0 g
Sugars: 10 g
Protein: 0 g

EXCHANGES PER SERVING

½ fruit

MAKES 8 SERVINGS | Preparation Time: 10 minutes | Chilling Time: 2 hours

peach-flavored green tea punch

Flavored with peach nectar, this punch is sure to refresh you and your guests.

6 cups water

6 peach-flavored green tea bags

3 orange-flavored black tea bags

2 cups peach nectar or unsweetened peach juice

⅔ cup SPLENDA® No Calorie Sweetener, Granulated

2 tablespoons fresh lemon juice

1. Bring the water to a boil in a medium saucepan over high heat. Remove from the heat. Add the green and black tea bags and cover. Let stand for 5 minutes. Remove the tea bags with a spoon, pressing the bags gently to release the flavor. Discard the tea bags.
2. Pour the tea into a punch bowl. Add the peach nectar, SPLENDA® Granulated Sweetener, and lemon juice and stir to dissolve the SPLENDA®. Cover and refrigerate until chilled, at least 2 hours or overnight. Serve in punch cups, over ice, if desired.

NUTRITIONAL INFORMATION

Serving Size: 8 fl oz
Calories: 35
Calories from Fat: 0
Total Fat: 0 g
Saturated Fat: 0 g
Cholesterol: 0 mg
Sodium: 10 mg
Total Carbohydrates: 9 g
Dietary Fiber: 0 g
Sugars: 8 g
Protein: 0 g

EXCHANGES PER SERVING

½ fruit

MAKES 12 SERVINGS | Preparation Time: 5 minutes | Chilling Time: 2 hours

pineapple-strawberry punch

Here's a colorful punch that will add zest to any party.

7 cups water

3 cups unsweetened pineapple juice

¾ cup SPLENDA® No Calorie Sweetener, Granulated

½ cup fresh lemon juice

One .14-ounce package Strawberry Flavor KOOL-AID® Unsweetened Soft Drink Mix

2 cups sliced fresh strawberries

1. Combine the water, pineapple juice, SPLENDA® Granulated Sweetener, lemon juice, and KOOL-AID® Unsweetened Soft Drink Mix in a large punch bowl. Stir well to dissolve the SPLENDA® and KOOL-AID®. Add the strawberries. Cover and refrigerate until chilled, at least 2 hours (or overnight).
2. Serve in punch cups over ice.

NUTRITIONAL INFORMATION

Serving Size: 8 fl oz
Calories: 45
Calories from Fat: 0
Total Fat: 0 g
Saturated Fat: 0 g
Cholesterol: 0 mg
Sodium: 20 mg
Total Carbohydrates: 11 g
Dietary Fiber: 1 g
Sugars: 9 g
Protein: 0g

EXCHANGES PER SERVING
½ fruit

MAKES 8 SERVINGS | Preparation Time: 5 minutes

sunshine punch

Citrus flavors are front and center in this classic punch—garnish with fresh fruit, if desired, for an especially festive touch.

2 cups ice-cold water

2 cups orange juice, chilled

1 cup SPLENDA® No Calorie Sweetener, Granulated

One .13-ounce package Tropical Fruit Flavor KOOL-AID® Unsweetened Soft Drink Mix

1 liter diet lemon-lime soda or diet ginger ale, chilled

Orange or pineapple slices for garnish (optional)

1. Stir the water, orange juice, SPLENDA® Granulated Sweetener, and KOOL-AID® Unsweetened Soft Drink Mix in a pitcher until the SPLENDA® and KOOL-AID® dissolve.
2. Just before serving, stir in the soda. Pour into glasses over ice. Garnish with orange or pineapple, if desired.

NUTRITIONAL INFORMATION

Serving Size: 8 fl oz
Calories: 30
Calories from Fat: 0
Total Fat: 0 g
Saturated Fat: 0 g
Cholesterol: 0 mg
Sodium: 25 mg
Total Carbohydrates: 7 g
Dietary Fiber: 0 g
Sugars: 6 g
Protein: 0 g

EXCHANGES PER SERVING

½ fruit

MAKES 11 SERVINGS | Preparation Time: 10 minutes | Chilling Time: 2 hours

raspberry tea punch

Fruit-flavored teas are a great flavor base for party punches, as shown by this raspberry-infused drink.

4 cups water

6 raspberry-flavored tea bags

2 orange pekoe tea bags

⅔ cup SPLENDA® No Calorie Sweetener, Granulated

1 liter club soda, chilled

¼ cup fresh lemon juice

2 cups fresh or frozen unsweetened raspberries

1. Bring the water to a boil in a medium saucepan. Remove from the heat. Add the raspberry and orange pekoe tea bags and cover. Let stand for 5 minutes. Remove the tea bags with a spoon, pressing the bags gently to release the flavor. Discard the tea bags. Add the SPLENDA® Granulated Sweetener and stir until it dissolves. Cover and refrigerate until chilled, at least 2 hours (or overnight).

2. Pour into a large pitcher. Stir in the club soda and lemon juice, then the raspberries. Pour into glasses over ice and serve.

NUTRITIONAL INFORMATION

Serving Size: 8 fl oz
Calories: 10
Calories from Fat: 0
Total Fat: 0 g
Saturated Fat: 0 g
Cholesterol: 0 mg
Sodium: 20 mg
Total Carbohydrates: 4 g
Dietary Fiber: 2 g
Sugars: 1 g
Protein: 0 g

EXCHANGES PER SERVING
Free

MAKES 8 SERVINGS | Preparation Time: 5 minutes

tropical pitcher punch

Mix up a pitcher of tasty punch with the tropical trio of pineapple, mango, and ginger.

½ cup SPLENDA® No Calorie Sweetener, Granulated

Two .13-ounce packages Tropical Punch Flavor KOOL-AID® Unsweetened Soft Drink Mix

2 cups unsweetened pineapple juice, chilled

2 cups mango nectar, chilled

1 liter diet ginger ale, chilled

1. Combine the SPLENDA® Granulated Sweetener and KOOL-AID® Unsweetened Soft Drink Mix in a large pitcher. Add the pineapple juice and mango nectar and stir well to dissolve the SPLENDA® and KOOL-AID®. Stir in the ginger ale.
2. Pour into glasses over ice and serve. Serve chilled.

NUTRITIONAL INFORMATION

Serving Size: 8 fl oz
Calories: 70
Calories from Fat: 0
Total Fat: 0 g
Saturated Fat: 0 g
Cholesterol: 0 mg
Sodium: 45 mg
Total Carbohydrates: 18 g
Dietary Fiber: 1 g
Sugars: 17 g
Protein: 0 g

EXCHANGES PER SERVING

1 fruit

tropical mango punch

Inspired by the flavors of the Caribbean, this crowd-pleasing punch features mango nectar and orange juice.

6 cups water

2 cups mango nectar, chilled

1 cup orange juice, preferably fresh, chilled

1 cup SPLENDA® No Calorie Sweetener, Granulated

One .13-ounce package Tropical Punch Flavor KOOL-AID® Unsweetened Soft Drink Mix

1. Combine the water, mango nectar, orange juice, SPLENDA® Granulated Sweetener, and KOOL-AID® Unsweetened Soft Drink Mix in a large pitcher or punch bowl. Stir to dissolve the SPLENDA® and KOOL-AID®.
2. Pour or ladle into glasses over ice and serve chilled.

NUTRITIONAL INFORMATION

Serving Size: 8 fl oz
Calories: 45
Calories from Fat: 0
Total Fat: 0 g
Saturated Fat: 0 g
Cholesterol: 0 mg
Sodium: 15 mg
Total Carbohydrates: 11 g
Dietary Fiber: 0 g
Sugars: 11 g
Protein: 0

EXCHANGES PER SERVING

½ fruit

MAKES 16 SERVINGS | Preparation Time: 10 minutes | Cooking Time: 30 minutes

mulled cider for a crowd

Enhance your winter holiday entertaining with fragrant mulled cider—keep it warm during the party in a slow cooker, if you wish.

- 8 cups unsweetened apple cider
- ½ cup SPLENDA® No Calorie Sweetener, Granulated
- 16 whole cloves
- 6 whole allspice berries
- Five 3-inch cinnamon sticks
- ⅓ cup fresh lemon juice
- ½ cup dried cranberries
- 8 thin orange slices
- 8 thin lemon slices

1. Combine the cider, SPLENDA® Granulated Sweetener, cloves, allspice, and cinnamon in a large nonreactive saucepan. Cook over very low heat, stirring occasionally, until hot, about 30 minutes. Do not boil. During the last 10 minutes, add the lemon juice, dried cranberries, and orange and lemon slices.
2. Ladle the cider into mugs and serve hot.

NUTRITIONAL INFORMATION

Serving Size: 4 fl oz
Calories: 80
Calories from Fat: 0
Total Fat: 0 g
Saturated Fat: 0 g
Cholesterol: 0 mg
Sodium: 15 mg
Total Carbohydrates: 19 g
Dietary Fiber: 0 g
Sugars: 16 g
Protein: 0 g

EXCHANGES PER SERVING
1 fruit

MAKES 12 SERVINGS | Preparation Time: 5 minutes | Cooking Time: 20 minutes

hot cranberry apple cider

A delicious variation on the mulled cider theme, this one has cranberry juice for added interest—be sure to use the unsweetened variety.

8 cups unsweetened apple cider
1 cup unsweetened cranberry juice
½ cup dried cranberries
Grated zest of 1 orange
⅓ cup orange juice, preferably fresh
1 cup SPLENDA® No Calorie Sweetener, Granulated
½ teaspoon whole cloves
½ teaspoon whole allspice berries
Four 3-inch cinnamon sticks

1. Combine the cider, cranberry juice, dried cranberries, orange zest, orange juice, SPLENDA® Granulated Sweetener, cloves, allspice, and cinnamon in a large nonreactive saucepan. Bring just to a simmer over medium heat. Reduce the heat to low and cook at a bare simmer for 20 minutes. Do not boil.
2. Ladle the cider into mugs and serve hot.

NUTRITIONAL INFORMATION

Serving Size: 6 fl oz
Calories: 100
Calories from Fat: 0
Total Fat: 0 g
Saturated Fat: 0 g
Cholesterol: 0 mg
Sodium: 15 mg
Total Carbohydrates: 25 g
Dietary Fiber: 1 g
Sugars: 21 g
Protein: 0 g

EXCHANGES PER SERVING
1½ fruits

ACKNOWLEDGMENTS

Just as a recipe needs the right blend of ingredients to taste its best, the perfect blend of people came together to create this wonderful companion piece. Our truest thanks and appreciation go out to everyone who dedicated their time and energy, especially the Copy Approval and SPLENDA® Brand teams. Thanks to the spirited, intelligent, and creative contributions of people like Diane, Desiree, Ed, Ivy, Janis, Liz, Marci, Mark, Matt, Maureen, Michelle, Rich, and Shideh; the final pages turned out as sweet as could be.

INDEX

A

Agua Fresca, Cantaloupe, 65
Almond Cappuccino, Frozen, 46
Apple cider
  Apple Breeze, 77
  Hot Cranberry Apple Cider, 117
  Mulled Cider for a Crowd, 114

B

Bananas
  Banana-Peanut Chocolate Smoothie, 11
  Banana-Raspberry Smoothie, 61
  "Bananas Foster," 58
  Fuzzy Orange Smoothie, 41

C

Cantaloupe Agua Fresca, 65
Cappuccino, Frozen Almond, 46
Chai
  Homemade Chai, 30
  Vanilla Chai Latte, 26
Chamomile-Pomegranate Iced Tea, 85
Cherry-Cranberry Warmer, Mulled, 29
Chocolate
  Banana-Peanut Chocolate Smoothie, 11
  Frozen Hot Chocolate, 19
  Hot Chocolate, 20
  Iced Mocha Latte, 15
  Instant Hot Cocoa, 16
  Malted Mocha Frappe, 57
Citrus Berry Spritzer, 74
Citrus-Mint Iced Tea, 86
Coffee
  Frozen Almond Cappuccino, 46
  Iced Mocha Latte, 15
  Malted Mocha Frappe, 57
  Warm Tiramisù Latte, 34
Cranberries
  Hot Cranberry Apple Cider, 117
  Mulled Cherry-Cranberry Warmer, 29
  Mulled Cider for a Crowd, 114
Creamsicle, 53

E

Eggnog
  chilling, 90
  Elegant Eggnog, 93

F

Fizzy Lemonade, 66
Float, Vanilla-Orange Yogurt, 78
Frappe, Malted Mocha, 57
French-Vanilla White Hot Chocolate, 12
Frozen Almond Cappuccino, 46
Frozen Hot Chocolate, 19
Fuzzy Orange Smoothie, 41

G

Ginger
  juice, 70
  Pear-Ginger Lemonade, 70
Ginger ale
  Orange-Berry Sparkler, 98
  Sunshine Punch, 106
  Tropical Pitcher Punch, 110
Grapefruit-Raspberry Sparkler, 69

H

Holiday Spiced Tea, 94
Homemade Chai, 30
Hot Brown Sugar Tea, 33
Hot Chocolate, 20
Hot Cranberry Apple Cider, 117
Hot Spiced Tea, 37
Hot Vanilla, 25

I

Ice cream
  Lemon-Lime Milk Shake, 45
  Malted Mocha Frappe, 57
Iced Mocha Latte, 15
Instant Hot Cocoa, 16

L

Lattes
  Iced Mocha Latte, 15
  Vanilla Chai Latte, 26
  Warm Tiramisù Latte, 34

Lemon-lime soda
Grapefruit-Raspberry Sparkler, 69
Lemon-Lime Milk Shake, 45
Orange-Berry Sparkler, 98
Raspberry Whip, 81
Sunshine Punch, 106
Lemons
Citrus-Mint Iced Tea, 86
Fizzy Lemonade, 66
Lemonade by the Pitcher!, 97
Lemonade Cubes, 54
Pear-Ginger Lemonade, 70
Refreshing Summer Slushie, 54
Watermelon Lemonade, 89

M

Malted Mocha Frappe, 57
Mangoes
Mango Yogurt Smoothie, 49
Tropical Mango Punch, 113
Tropical Pitcher Punch, 110
Mulled Cherry-Cranberry Warmer, 29
Mulled Cider for a Crowd, 114

O

Oranges
Citrus Berry Spritzer, 74
Citrus-Mint Iced Tea, 86
Creamsicle, 53
Fuzzy Orange Smoothie, 41
Holiday Spiced Tea, 94
Orange-Berry Sparkler, 98
Pomegranate Punch, 73
Sunshine Punch, 106
Tropical Mango Punch, 113
Vanilla-Orange Yogurt Float, 78

P

Paradise Punch, 105
Peaches
Fuzzy Orange Smoothie, 41
Peach-Flavored Green Tea Punch, 102
Peach Melba Sipper, 50
Peanut Chocolate Smoothie, Banana-, 11
Pear-Ginger Lemonade, 70
Pineapple
Paradise Punch, 101
Pineapple-Strawberry Punch, 105
Tropical Pitcher Punch, 110
Pomegranates
Chamomile-Pomegranate Iced Tea, 85
Pomegranate Punch, 73
Punches
chilling, 90
Orange-Berry Sparkler, 98
Paradise Punch, 101
Peach-Flavored Green Tea Punch, 102
Pineapple-Strawberry Punch, 105
Pomegranate Punch, 73
Raspberry Tea Punch, 109
Sunshine Punch, 106
Tropical Mango Punch, 113
Tropical Pitcher Punch, 110

R

Raspberries
Banana-Raspberry Smoothie, 61
Citrus Berry Spritzer, 74
Grapefruit-Raspberry Sparkler, 69
Orange-Berry Sparkler, 98
Peach Melba Sipper, 50
Raspberry Sauce, 50
Raspberry Tea Punch, 109
Raspberry Whip, 81
Refreshing Summer Slushie, 54

S

Shakes
Lemon-Lime Milk Shake, 45
Malted Mocha Frappe, 57
Slushie, Refreshing Summer, 54
Smoothies
Banana-Peanut Chocolate Smoothie, 11
Banana-Raspberry Smoothie, 61
"Bananas Foster," 58
Fuzzy Orange Smoothie, 41
Mango Yogurt Smoothie, 49
Peach Melba Sipper, 50
Strawberry Smoothie, 42
Southern Iced Tea, 82
SPLENDA® Sweetener Products
substitution chart for, 7
types of, 6
Strawberries
Pineapple-Strawberry Punch, 105
Strawberry Smoothie, 42
Sunshine Punch, 106

T

Tea
Chamomile-Pomegranate Iced Tea, 85
Citrus-Mint Iced Tea, 86
Holiday Spiced Tea, 94
Homemade Chai, 30
Hot Brown Sugar Tea, 33
Hot Spiced Tea, 37
Peach-Flavored Green Tea Punch, 102
Raspberry Tea Punch, 109
Southern Iced Tea, 82
Vanilla Chai Latte, 26
Tiramisù Latte, Warm, 34
Tropical Mango Punch, 113
Tropical Pitcher Punch, 110

V

Vanilla
French-Vanilla White Hot Chocolate, 12
Hot Vanilla, 25
Vanilla Chai Latte, 26
Vanilla-Orange Yogurt Float, 78

W

Warm Tiramisù Latte, 34
Watermelon Lemonade, 89
White Hot Chocolate, French-Vanilla, 12

Y

Yogurt
Mango Yogurt Smoothie, 49
Peach Melba Sipper, 50
Strawberry Smoothie, 42
Vanilla-Orange Yogurt Float, 78

table of equivalents

Bar spoon	=	½ ounce
1 teaspoon	=	1/6 ounce
1 tablespoon	=	½ ounce
2 tablespoons (pony)	=	1 ounce
3 tablespoons (jigger)	=	1½ ounces
¼ cup	=	2 ounces
⅓ cup	=	3 ounces
½ cup	=	4 ounces
⅔ cup	=	5 ounces
¾ cup	=	6 ounces
1 cup	=	8 ounces
1 pint	=	16 ounces
1 quart	=	32 ounces
750 ml bottle	=	25.4 ounces
1 liter bottle	=	33.8 ounces
1 medium lemon	=	3 tablespoons juice
1 medium lime	=	2 tablespoons juice
1 medium orange	=	⅓ cup juice